The Homer Spit

Coal, Gold and Con Men

Janet R. Klein

Janet Klein

A Centennial Publication
1896-1996

Printed in the United States of America.
ISBN # 0–9651157–1–2

Klein, Janet R.
Homer Spit
Homer, Alaska
Kachemak Bay

Front Cover:
Photograph – Homer Pennock, 1898
Map – Russian map of Kenai Peninsula, 1852

Funded in part by the City of Homer
in honor of the 1896-1996 Centennial Celebration

Layout and Design by:
Michael R. Ward, Hard Luck Mines, Homer AK

Contents

Preface

This booklet was compiled to celebrate the Centennial of the naming of Homer, Alaska in 1896. Primarily, it's a compendium of quotations and illustrations from the earliest recorded history of the Homer Spit through the early 1900s. To retain the flavor and character of those decades, the original spellings of words have been retained along with occasional inaccuracies in information. The inaccuracies, however, are corrected elsewhere in the text. As one who has searched for 17 years for a photograph of our City's founding father, I harbor a slight skepticism that the photograph clearly identified as Homer Pennock is, indeed, him. Possibly it was a set up, another con job he perpetrated just to appease a photographer or to perplex the public!

Jon Faulkner, Homer historian and owner of Land's End, wrote in the acknowledgement to an exhibit about the Homer Spit, "History, in its many forms, is our collective heritage. Similar to our individual lineage, a community's history defines its roots, who we are as a social unit, how we look and function, what values we embrace, and, thus, how our future is shaped.

"How often do we look back at earlier times and comment, "Wow! If only I had lived then." What is it about then that often makes it so alluring? Is it our interest in people? Perhaps our fascination with the hardships they endured? Is it our attraction to a simpler lifestyle, or to an unexplored, untouched wilderness?" This booklet offers a glimpse of those times and of those people who saw the Homer Spit when it was still relatively wild and unaltered.

The grandsons of Della and Austin Banks generously provided information about their grandparents and allowed me to reproduce excerpts from her articles. Hopefully, this publication will offer them fresh insights into their grandparents' Northern adventures and the important role Della, in particular, played in documenting our City's beginnings. The City also helped support this project as part of its contribution to the Centennial Celebration of the naming of Homer.

"I'd sure give my eye teeth to see what the Spit looked like 500 years ago."

Joel Moss, Retired Homer fisherman
who first saw the Homer Spit in 1946

Introduction

Kachemak Bay is a magnificent embayment angling off the lower end of Cook Inlet in Southcentral Alaska. It slices about 40 miles into the southern end of the Kenai Peninsula, separating the younger, rolling hills of the western Peninsula from the older glaciated and rugged 4,000+ foot peaks of the Kenai Mountains. The funnel–shaped Bay narrows from about 25 miles at its mouth to about six miles at the Fox River Flats.

The Homer Spit is one of the most striking, unique features within Kachemak Bay. This natural finger of gravel and sand stretches four and one–half miles into the Bay. Remnant ridges of native timothy and flowers such as beach peas and lupine, struggle for survival above the highest tide lines and in the lee of wind–buffeted sand berms. Spruce, which once forested the Spit, are repopulating the roadside.

The Spit, as it's known locally, separates the waters of Kachemak Bay into two ecosystems – an inner Bay and an outer Bay. Each is geographically and biologically unique. Inner Bay waters extend to the head of Kachemak Bay where a tangle of fresh and glacial waterways undulate across the Fox River Flats. Nutrient–rich sediments are washed from the great bluffs rising above the beaches along the north shore or are washed down Bay from Sheep Creek and the Bradley, Fox and Martin Rivers to be deposited in a mile–wide arc of flats fringing the shore. Large, isolated boulders, known as glacial erratics, dot the flats which diminish rapidly and almost disappear at the tip of the Spit.

Outer Kachemak Bay includes the deeper, saltier waters south and west of the Spit which open directly into lower Cook Inlet. Archimandritof Shoals parallel the landmass on the west. Longshore currents, flowing along the Spit, transport sediments and detritus to its beaches.

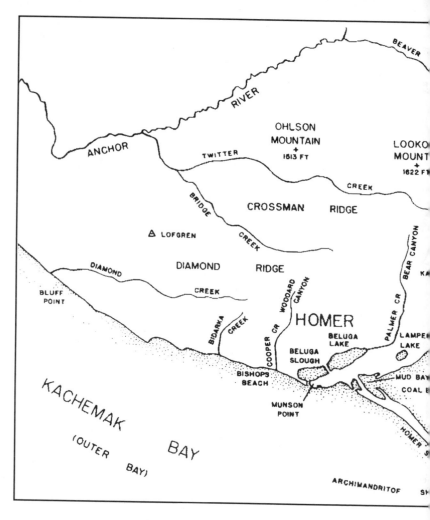

Although Homer was located on the tip of the Spit from 1896 to 1902, contemporary Homer, at its base, began about 1917. Coal in the

Prehistoric People Visited the Spit

Time and distance obscure any records of the very first visitors to the Spit; however, at some point Native Alaskans utilized the landmark. Near the base and about three–fourths of the way out

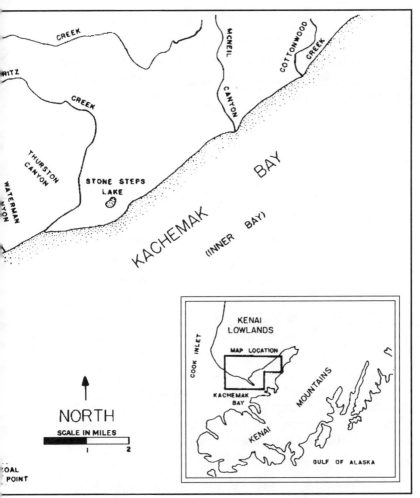

from *A History of Kachemak Bay, the Country, the Communities*
soaring sandstone bluffs between Bidarka Creek and Cottonwood Creek was the focus of intense mining a century ago.

were two middens. These concentrations of shells, fire–cracked rock and a few tools indicate that the prehistoric visitors built fires and probably gathered and ate the clams and mussels whose shells they discarded. One midden was destroyed during the construction of the Homer Small Boat Harbor. The other has been eroded by tidal action for decades and only a trace remains. Dena'ina Athabaskan

Indians, who probably visited Kachemak Bay about 1,000 years ago, were familiar with the Spit. Peter Kalifornsky was a revered elder and author who recorded much of his peoples' history. He recalled that hunters gathered at the base of the Spit before going sea otter hunting. *Uzintin*, the Dena'ina word for the Spit, means "extends out into the distance." *Tikakaq*, or Kachemak Bay, means "ocean mouth."

British Navigators By-passed Kachemak Bay

Date	Captain	Ship
1778	James Cook	Resolution
1786	Nathaniel Portlock	King George
	George Dixon	Queen Charlotte
1786–88	John Meares	Iphigenia
	William Douglas	
1794	George Vancouver	Discovery

British navigators charted much of the coast of Cook Inlet. Subtle signs in the tides, currents or the lay of the land must have indicated to these great seamen that Kachemak Bay was just a smaller embayment off of the Inlet and not especially worth investigating.

Russian Navigators Explored Kachemak Bay

The first non–Native place names for Kachemak Bay and for the Spit were recorded by Captain Mikhail Tebenkov. As director of the Russian American Company and governor of Russian America (Alaska) from 1845–50, he surveyed much of Alaska's coast. His map may well be the first to depict the Spit.

Between 1830–1850, Chernof, Dinglestadt, Doroshin and Archimandritof, employees of the Russian American Company, explored parts of the Kenai Peninsula. Sketchy records suggest that several of them explored Kachemak Bay. On Vosnesenski's map,

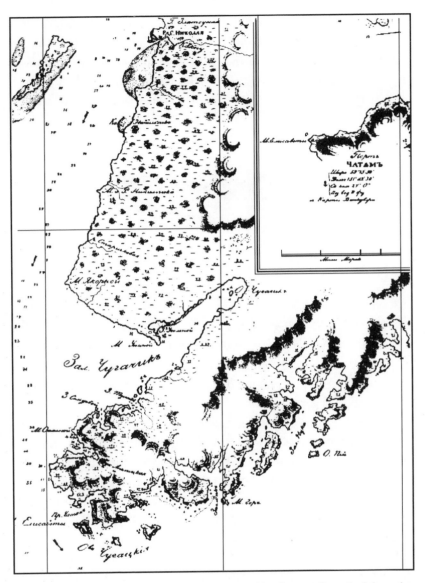

from *Atlas of the Northwest Coast of America*

Tebenkov's map, published in 1852, may be the first to illustrate the Spit and local place names: *Mys Iakornyi* (Anchor Point), *Z. Ugolnoi* (Coal Bay), *Mys Ugolnoi* (Coal Point) and *Z. Chugachik* (Chugachik Bay). Note: *Ugolnoi,* when used as an adjective, can mean either 'coal' or 'corner'. Because the Russians were probably documenting local resources, it is assumed that *Ugolnoi* means 'coal.'

published in 1850, Kachemak Bay is identified as *B. Tschugatschik oder Kotschekmaksky*. Both names were probably Russianized Native Alaskan words. The lack of contours along the shores of inner Kachemak Bay suggest that the coast was not surveyed.

Russian entrepreneurs operated a coal mine in Port Graham between 1855–1867; however, no record exists to show if they visited the Spit.

The Late 1800s
The Advent of the Americans

In 1867, the United States purchased Alaska from Russia for $7.2 million. American explorers and scientists, primarily working for the federal government, spent decades investigating and documenting the geology, plant and animal life, waterways and cultures of the 586,000 acre territory.

Anchoring just inside the tip of the Homer Spit, during these years, were ships from the U.S. Revenue–Cutter Service, the U.S. Bureau of Fisheries, the U.S. Coast and Geodetic Survey and the U.S. Geological Survey. Aboard them were cartographers, geologists, ornithologists, glaciologists, artists and others. Military expeditions also occasionally regrouped in Kachemak Bay.

The earliest description of the unnamed, unoccupied Spit recorded by Americans dates to 1880 with the visit of William H. Dall and Tarleton H. Bean aboard the schooner *Yukon*. Dall, then Acting Assistant for the U.S. Coast and Geodetic Survey, mapped the coast and bestowed place names on many features. Bean, who was investigating the fish and fisheries of Alaska, also collected bird specimens for the National Museum (Smithsonian Institution).

Bean's description of the Spit hints of its former beauty and bounty.

William Healy Dall, a Massachusetts scientist shown here in 1879, spent over 30 seasons studying the resources of Alaska. He visited Kachemak Bay in 1880, 1895 and 1899 to chart the coastline and document the geology and glaciers.

from the Smithsonian Institution Archives, SA 1145

"The spit in Chugachik (Kachemak) Bay, on which I collected birds July 1, 1880, is low and level, its beaches higher than the interior. At some high tides, the sea breaks over and carries with it immense numbers of fish, which are left stranded when the waters recede. This occurred a few days before our visit, and we saw thousands upon thousands of fishes lying uncovered on the ground. Great quantities of drift-wood are found here. Wild wheat abounds, and there are many pretty flowering plants, among which are serrana (chocolate lily), violets, chickweed, vetch, and Jacob's ladder. There is a little grove of Sitka spruces, in which I found the redpolls and thrushes. On this spit was found the young eider which I have numbered in my catalogue."

"On the 2nd of July we visited Glacier spit, distant 9 miles from

our anchorage. Here a pair of eagles had a nest on one of the tall pines."

Ivan Petroff, who conducted the first census in Alaska, was also here in 1880 and recorded 74 inhabitants in Seldovia and Ostrovsky. How far Petroff ventured into Kachemak Bay in his search for settlements remains unknown; however, neither he, Dall nor Bean mentioned people living on the Spit.

1883

Rising on the horizon about 60 miles west of the Spit is Mt. Augustine Island, also a volcano. In 1883, the classically shaped cone erupted violently.

A tsunami, estimated to be about 30 feet high, flooded into the communities of Port Graham and Fort Alexander (today, Nanwalek). Fortunately, the towering wall of water arrived during a low tide so it caused minimal damage.

In all likelihood the great wave crested as it encountered the Archimandritof Shoals, then crashed over the Spit. Undoubtedly, it redistributed the sediments and reshaped the contours of the fragile finger of land.

The Catalyst — Coal

Coal was the catalyst fueling the first mineral explorations and extractions in Kachemak Bay. The Kenai Formation, underlying the north shore, is composed of coal–bearing clays and oil and gas–bearing sands. Beds of coal outcrop in the sandstone bluffs that stretch from the Fox River Flats westward and northward to Clam Gulch.

The quality of coal varies. Geologists have identified lignite, bituminous and sub–bituminous outcrops. Early geological reports stated that the lower, deeper strata were of better quality than the higher. Most of the coal seams are two to three feet thick yet a few reach six or seven feet. Tunnels were driven into these thicker seams.

Bunkers, or storage bins, near the Alaska Coal Company store, were used to stockpile coal. The distant, high mountaintop known unofficially today as Sadie Peak, was called Tyndal Peak in 1892.

Exactly when the first prospecting and mining of coal occurred on the north shore remains unknown. About 1888 the Alaska Coal Company tunneled into the Bradley seam which suggests that the Company was here prior to that year. Coal Bay, a townsite of almost 1,280 acres, was claimed by the ACC and the post office operated from August 8th, 1892 to June 21, 1895.

The 1890s

Increasing numbers of ships and people were attracted to Cook Inlet in the 1890s. Gold, found on Inlet beaches, on Turnagain Arm

9

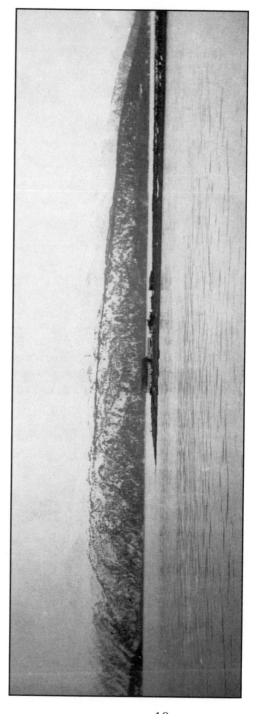

from National Archives, 22–FFA–484

The earliest photographs of buildings on the Spit date to 1892. The U.S. Fish and Fisheries Commission steamer *Albatross* anchored offshore and crewman, "NBM," photographed the tiny settlement called Station Coal Point.

The population of Station Coal Point was eleven white male residents in 1892. The Native men, posed near a sawhorse and tools alongside the Alaska Coal Company store, were probably daily workers.

and along numerous rivers, precipitated a minor rush from 1895 to 1897, and moose, Dall sheep, mountain goat, bear and other animals attracted big game hunters and museum collectors to the Peninsula toward the end of the decade. Coal was used to heat cabins, cook food and fuel ships, especially the fleet of smaller boats transporting freight, miners, hunters and mail to the upper Inlet.

Kachemak coal occurs in several grades and samples were tested sporadically over the decades. In 1891, Lieutenant R. P. Schwerin of the U.S. Navy collected samples from four localities. Tested in San Francisco, all was found to spark excessively which posed a fire hazard to trains crossing the dry California mountains. Few commercial markets were ever developed for Kachemak coal. Most was used in Southcentral Alaska or by steam ships running from Southeast Alaska to the Aleutian Islands.

The quality of the coal, however, did not deter companies from prospecting, promoting and producing some quantities of fuel. Although the mining occurred at the base of the Spit and near McNeil, Cottonwood and Eastland Creeks, the tip of the Spit was headquarters for numerous companies over the years.

11

The Spit was surveyed numerous times. During August and September, 1892, Albert Lascy mapped its perimeter. His drawings revealed that the Spit was once forested as he indicated areas of "dead timber" and "live timber." Evidence suggests that most spruce had died and fallen decades earlier as the Sitka surveyor noted only two groves of living trees.

The plat of survey #105 includes the tip of the Spit. Five structures, including two cabins, coal bunkers, the Alaska Coal Company store and the Cooper Coal and Commercial Company Store, cluster on the inner end. The Cooper store may have been owned or operated by Joseph Cooper who mined on the Kenai Peninsula for many years.

After a 15 year absence, William H. Dall returned to Kachemak Bay in 1895. Now with the U.S. Geological Survey, he found the North Pacific Mining and Transportation Company and the Alaska Coal Company actively tunneling for the black fuel in upper Kachemak Bay. Near McNeil Canyon they had dug several tunnels and built a wharf and a bunker. These companies had abandoned Coal Bay and

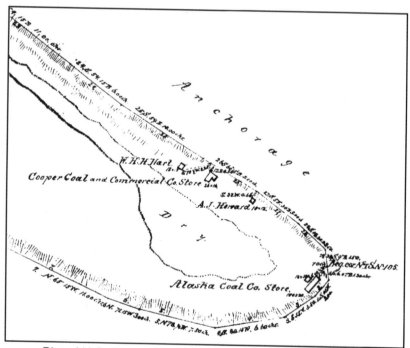

Plat of U.S. Survey No. 105. Surveyed August 30 and 31, 1892.

12

their buildings on the Spit. Dall wrote,

"I received the impression that these two company corporations, if not composed of the same members throughout, were at least not competitors, and were sustained chiefly by capital from a common source."

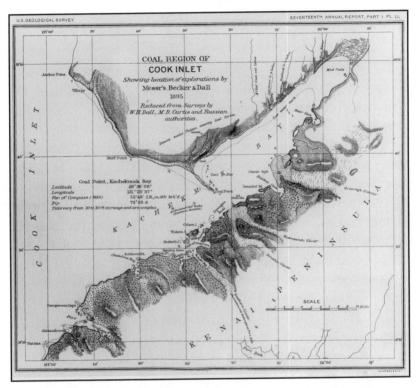

courtesy of the U.S. Geological Survey, 17th Annual Report
Detailed land formations on this 1895 map were surveyed by many map makers. Dall bestowed most of the place names on features during his first trip to Kachemak Bay in 1880. Note that Aurora Lagoon opens to the south, Neptune Bay does not exist yet and China Poot Bay is filled with tidal flats.

Dall suggested that a railroad could be built from the bluffs at the base of the Spit to the tip where a natural deepwater anchorage existed. Large ships could take on coal without having to regulate their schedules around the tides of Kachemak Bay. In 1899 that railroad was built.

Homer Pennock

Who was this man for whom our community was named a century ago? What is known of him that would support the sentence, from *An Adventure in Alaska,* that he was "... the most talented confidence man that ever operated on this continent"?

Scant information about Homer Pennock exists. Terse references to him and his operations occur in mining records, magazines, newspapers, books, city directories, census records and on his death certificate. Contradictory information exists about Pennock, sometimes even within one article. Some mistakes are relatively insignificant; others may have been perpetrated or left uncorrected by Pennock, himself.

Con men often prefer to remain secretive, elusive, anonymous; and, erroneous information can be beneficial. For example, Della Murray Banks and her husband, Austin, accompanied Pennock to the Spit in 1896 and 1897. In a series of articles describing their experiences, she called him a "Michigan man." Pennock is a surname common in that state, there's a town called Homer (not named after this man), and his first swindle was spearheaded out of the Detroit area; however, Pennock was a New York man. He was born there, he maintained a working residence there much of his life and he died there.

The Otter Head Tin Swindle

The earliest reference to Homer Pennock and, possibly, the first scam he concocted occurred in Canada in 1871. The 31 year old "discovered" an incredible outcrop of tin. An elaborate report, complete with a geologic map and schematics of the outcrop, was published. Pennock and partners induced American capitalists around Detroit to form an investment company. Surveyors mapped an entire township, named Homer (presumably after him), which extended about 12 miles along the coast and 10 miles inland from the Canadian shore of Lake Superior.

A curious mining geologist who questioned the presence of such rich tin in the region, secretly explored the site and found it fraudulent. He exposed the scam yet investors ignored his warning and continued to purchase stock.

In 1873 when rumors of the swindle resurfaced, investors listened. Pennock was caught, convicted of having the mines salted with tin procured in Cornwall, England and jailed in The Tombs in New York City.

The Otter Head Tin Swindle revealed one type of scam Pennock would perpetrate across North America. If legitimate means of securing fortune failed, he devised alternative ways of mining money, often by floating bogus paper stock on the market.

Pennock must have been persuasive, possibly even charismatic, for he continually found financial supporters for his mining ventures.

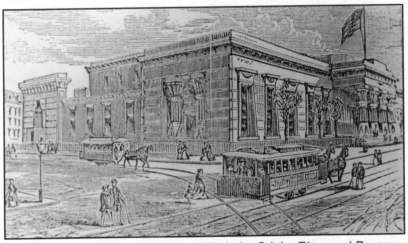

from *History of the City of New York, Its Origin, Rise, and Progress*
The Tombs, where Pennock was jailed, was described as "... the purest specimen of Egyptian architecture to be found outside of Egypt itself."

15

from the Yukon Archives/MacBride Museum Collection, print #3661
**Homer Pennock avoided photographers in Kachemak Bay in 1896
and 1897 yet was caught by a camera two years later as he hunted
for gold in the Klondike. This, the only known photograph of him,
shows the con man at age 58, then a member of The Mysterious 36,
a secretive group which, like so many others, failed to find gold in
the Klondike.**

Fortune at Forty

By 1877, Pennock was out of prison and in Leadville, Colorado. Little induced him to remain in the rugged, youthful silver camp high in the Rocky Mountains so he traveled westward. Two years later he returned to Leadville.

Pennock was almost forty. Supported with money from East Coast investors, he and several partners gambled on expanding an abandoned mine. As a co–owner and engineer of the project, Pennock prodded the miners deeper into the tunnels where they finally struck the silver lode. The Robert E. Lee mine took the front rank as a producer from the start, a position it maintained for several years. During 1880 over one million dollars was divided among the stockholders.

A personal history of Pennock and one partner, L.D. Roudebush, states,

"These gentlemen have invested boldly and successfully in Leadville mining properties... Mr. Pennock has spent twenty years in mining regions, from Lake Superior to California and Nevada. Following every excitement, he has experienced the ups and downs of such a life, is thoroughly conversant with ores, and experienced in all the surface indications and underground requirements of what should constitute a promising prospect or a well opened mine.

Years of thoughtful observation and the memories of many glittering failures have taught him that wisdom which precludes visionary conclusions at sight of every fine mineral showing, but when there are substantial grounds for belief in the permanence and richness of a deposit of ore, his practiced eye and experienced judgment grasp these salient features and the worth of a property is quickly arrived at. His first visit to Leadville in 1877, when the bulk of ores was low in silver and high in lead forced the belief that the camp was not a paying one, but later, when development brought forth ores of a richer grade in silver, another examination convinced him of the immense value of the carbonate leads, and he at once reversed his decision. At this time a letter from T.W.B. Hughes, a member of the New York Stock Exchange, was received,

17

asking for information regarding the mines of Leadville. Pennock's reply counselled Hughes to sell his seat in the Board and bring the money ($5,000) to Colorado for investment in the carbonate mines, closing with the remark: "This is a sure opportunity to make a fortune with very little risk." Hughes showed this letter to L.D. Roudebush, also a member of the Stock Exchange, and Roudebush having known Pennock for thirteen years, and possessing unbounded confidence in his integrity and judgment as a practical mining expert, obtained Hughes' ready assent to his action upon the information... ...and without any previous experience in mining or knowledge of mining operations, he started to Colorado to "make or break," under the advice and co-operation of his friend and present partner, Homer Pennock."

The author concluded,

"...and Pennock can speak of the rough scenes of his orphan boyhood and hardy prospecting career as contrasting beautifully with the present smiles of dame fortune, for both partners are ranked among the millionaires of the carbonate metropolis."

from *Leadville*

Corsets, Car Wheels and the Collapse of a Fortune

After amassing millions in Colorado, Pennock invested in a corset factory in Detroit then an elastic steel car wheel plant in Chicago. He was listed in the Chicago city directories from 1883 to 1885 as president of the Miltimore Company and of the Western Brick and Tile Manufacturing Company. He also owned a large subdivision, Pennock Village.

"This village is owned by Homer Pennock, who has recently purchased about twelve hundred acres of land in the immediate vicinity, upon which is located the buildings of the Miltimore Elastic Steel Car Wheel Company, where the celebrated elastic steel car wheels invented by John Miltimore are made."

from *History of Cook County, Illinois*

18

He *"...was worth about $4 million when he tired of this activity and went down into the Transvaal to dig diamonds. It was there that he lost the greater part of his fortune."*

from *the Denver Post*

What happened in the intervening years remains unknown. By 1896, Pennock appeared penniless and back on the mining circuit. A minor gold rush in Cook Inlet captured his attention and, as manager of the Alaska Gold Mining Company, he brought a crew North.

Della Murray Banks

On March 18, 1896 a determined young woman, Della Murray Banks, and her husband departed Seattle aboard the *Excelsior* for Kachemak Bay. She and Austin, residents of Colorado, had suffered financially in the panic of 1893 and the promise of gold glittering in the beach sands of Cook Inlet and in their pockets enticed Austin Banks to join the Alaska Gold Mining Company managed by Homer Pennock. That first summer stretched into five as they followed the golden circuit from Alaska to Canada.

During those seasons, city–born and bred Della Banks met and bested the challenges of her new life. After her sink–or–swim initiation into country living, she became tough and tempered. Although only five feet, one inch tall and about 108 pounds, the indomitable brunette became comfortable in the mining camps of Alaska and Canada.

After her last trip North in 1901, Della, Austin and Sydney, their young son, settled in Seattle to be close to the Northland. Austin Banks died in 1919 and Della and Sydney moved to Los Angeles

County, California, yet the urge to explore never left her. On a trip into the Kentucky hills, she fell from a horse. Paralyzed and confined to a wheelchair until her death, she discovered new joys and sustenance in her memories. She devoured news of the people she had met and of the country she had explored.

Almost 50 years after arriving on the sandspit in Kachemak Bay, she laboriously typed a series of articles illustrated with her photographs. They describe the irresistible lure of gold–hunting one hundred years ago, yet more than that, they chronicle her changing from a city woman to a plucky pioneer who found her fortune, finally, in the friendships, memories and pictures of the North.

Della Banks called herself "Homer's first historian." Although once she described the Homer Spit, "...as desolate a spot as could be imagined," time, distance and reflection replaced that initial feeling. Memories of the land and the people "...never disappoint you as a dream of fortune does," she wrote late in life.

Irrepressible and tough, Della Banks met the challenges of the North and became as resilient and spirited as the land she learned to love.

Homer's Gold Seekers

"Present–day Homer straggles for four or five miles along the low bluff running back to the rolling hills on the north shore of Kachemak Bay, in lower Cook Inlet. A town of from two to three hundred people with homes, wonderful flower and vegetable gardens, schools... that is Homer in 1945.

But the old Homer, the Homer I knew in 1896, was one log house, two tumbledown shacks, and the galley of some ill–fated ship, clustered haphazardly at the outer end of the long sand–bar bordered spit which juts out like a long finger from the bluff into the bay.

Homer didn't even have a name on that first day of April, 1896–April Fool's Day, how prophetic! In making application for a post office, we had to have a name. With the rest, I voted for the name

"Homer," in honor of Homer Pennock, the mining promoter and manager of the Alaska Gold Mining Company with which we were associated.

But perhaps I should begin with the story of how I happened to go on that first wild–goose chase in search of a fortune in gold. My husband, Austin Banks, and I were living in Denver. Pennock came through on his way west, enthusiastic over his "sure–fire discovery

A year before her first Alaskan adventure, Della Murray Banks posed for this studio shot in Denver, Colorado. Born and raised in big cities, the spirited 31 year old insisted on accompanying her husband to the gold fields of Alaska.

photograph by Della Murray Banks, 1898

All the other men's wives wanted to go along to Cook Inlet, but I simply went. For a city–bred woman who had never even camped out overnight before, life on the Inlet required a bit of adjustment.

of gold in Alaska." Austin had known Pennock for years, and listened eagerly to his story of certain wealth; for, like so many others, Austin had lost everything in the panic of '93. We had seen hard times since our marriage, in the late fall of the panic year.

Pennock was taking half a dozen Denver men with him, and, to my great astonishment, Austin was set on going, too. It would be "only for the summer"; they would return in September, well paid for the summer's work in the then–unknown country. I had been a proofreader on the Denver Times, and had worked on at the office for several months after our marriage. But now I was surprised at Austin's casual idea that I could go back to work while he was gone.

When I found out that he was really in earnest... and that I was supposedly keeping him from making a fortune, I agreed to his going. They left Denver on March 10. Four days later I wired him to come back, for I would not stay there without him. He wired back that I should take the next train for Seattle; the boat would sail on the following Tuesday. His wire reached me at eight–thirty Saturday night. Four hours later I was on the train, leaving my house for my sister to rent for the six months I should be away.

It was twenty–five years before I saw Denver again, and then I was a widow.

Austin met the train Monday night. We took the morning boat for Tacoma, where the *Excelsior*, a steam schooner Pennock had chartered, was loading. She wouldn't get underway before Thursday. Pennock was up town, so we walked up the half–mile incline from the old wharf and met him.

Pennock was more than six feet tall, must have weighed two hundred and fifty pounds, and had a grizzled beard and hair. His gray eyes were hard, keen, and shrewd. He looked down at me, five feet and one inch tall, and at best a hundred and eight pounds.

"So you think you have to go, too, do you?"

I looked up at the grim face and gathered my courage.

"No, I don't have to go. But if Austin goes, then I have to go."

Pennock laughed. "Then you have to go, for I must have Austin."

So it was settled. Pennock said later that all the other men's wives

23

had asked to go, but that Mrs. Banks hadn't even asked. She just went. But Pennock felt sure that when I saw my accommodations for the trip, I'd decide to stay home. There was a cabin, which was also the dining room; a small galley; eight staterooms with two or four bunks each, and only four of those staterooms available for passengers. Certainly none was to be given over to a rather unwelcome lady!

There were seventy–five men going, most of them for the Company. They were to be quartered in the hold. First, layers of lumber. Then bunks in three tiers. Boxes, bales, and heaps of freight were stack all about. The entire main deck was loaded with lumber. On top of it, stalls had been built for the twenty–one horses; and sacks of coal were piled on top of the stalls.

* * *

Friday morning we steamed up to the sandspit. There it was, our destination. Across Kachemak Bay were snow–covered mountains with glaciers winding between them. Behind the sandspit the bluff was covered with small evergreen timber, and snow lay on the ground. The edges of the spit were piled high with slabs of ice. There was no smoke, no sign of life at the lone log cabin, and the blast of the steamer's whistle roused no one.

Guilbault said the caretaker must be at the coal camp, so we sailed on ten miles to McNeill's Canyon, where a coal mine was being worked. We learned then that Edmond Guilbault...had made two trips to Kachemak Bay in 1894 and '95. He had been at the coal camp a year, and it was through him that Pennock had become interested.

We picked up Fritz, the caretaker, and went back to the sandspit, reaching the anchorage at low tide when there was a band of sand showing between the water and the ice. ...everything had to be brought in by rowboat, carried up the steep bank, and stored in the old house.

The next day Austin and I settled in the old ship's galley... We had plenty of fuel–the entire spit was covered with broken coal washed up on the shore from the seams in the bluffs.

Coal seams, lacing the sandstone bluffs of the north shore, range in thickness from mere inches to about seven feet. Tunnels were bored into a seam at Eastland Creek in the 1890s and into this thick bed near Bidarka Creek in 1899.

photograph by Janet Klein

It (the spit) was as desolate a spot as could be imagined; and one thing stood out clearly–the place was as new and strange to Pennock as to any of us. I remembered his hesitant answer when I had asked him if he had been there. One day I said to Young Ed, "Pennock has been here before, hasn't he? He told us on the boat that he had."

"No, he hasn't," admitted Young Ed. "But he had to tell the eastern men he had seen it. It he hadn't, they wouldn't have come."

That was logical and probably true, if not ethical; but ever after I doubted Pennock.

A bunk house was built, and the two shacks were repaired. The log house was about a hundred by twenty feet, divided both upstairs and down into five equal rooms.

When the *Excelsior* sailed south a week later she took most of the independent gold–seekers with her, as well as a few malcontents and trouble–makers of our own group.

Until the ice melted away, we could go no farther up the Inlet for prospecting. We gathered coal, explored the barren sandspit back to the bluffs, and, climbing over the rough ice barrier, walked the broad sand bars at low tide. We played whist, melted ice for water, watched the smoke of Iliamna Volcano, and saw the mirage that makes the cone of Augustine Volcano square as a cube and Redoubt and Iliamna appear to be hour glasses. The scenery was gorgeous; but man does not go into unknown lands for scenery alone, so the men at camp grumbled at the late spring and talked of the sure fortune soon to be theirs.

Before the middle of May, the *Lakme* and the *Utopia* arrived with four hundred men headed for the Turnagain Arm district. They

courtesy of Mona Painter

Joseph Cooper, second from the left, back row, explored much of the Kenai Peninsula in his search for gold. Although he lived in Ninilchik, Cooper Landing is named for him and he had a cabin at the base of the Spit.

camped on the spit, and I soon learned that the free, wide–open spaces weren't quite free to me.

Among those on the *Lakme* was Soapy Smith, come up to see what would be the prospects there; but the pickings were going to be too small for a genius like Soapy, and he went back on the same boat looking for greener pastures. He found them the next year in Skagway.

It was on the sandspit that I got my first glimpse of the gold we had come so far to find. I was standing in the little cabin door when a black–bearded man came up and introduced himself as Cooper from Ninilchik. He hoped I would get up to meet his wife during the summer. A woman neighbor only forty miles away! Then he opened a small package.

"For luck to a new–comer to our country," he said, handing me a piece of amalgamated gold. It was a kind, neighborly gesture, and I was a very homesick woman!

When the ice began to melt away, the question of fresh water became paramount. Two lighters had been built, one a hundred feet long and forty feet wide, to be used for carrying supplies up the Inlet. But when it was completed, it was found impracticable because of its size and the tides. It was torn apart, and one small enough to be towed by a rowboat was built.

Loaded with barrels, this lighter was towed to the bluff behind the spit, where a small stream flowed into the bay. It was only ten miles round–trip, and it took all day. Next time we crossed to Halibut Cove for water.

The *General Canby,* a tugboat Pennock had bought in Tacoma, arrived early in May.

The *Canby* took a crew of men and supplies up to Anchor Point, about twenty–five miles up the Inlet, where houses were to be built and placer–mining operations started along the beach sands. Austin went with them. Another crew went to Snug Harbor, across the Inlet, while a third went up to Resurrection Creek, in Turnagain Arm.

Then the *Canby* came down from Anchor Point, and I leaned that she was going up to Hope, the new town on Resurrection Creek in

Turnagain Arm, to see the men we had sent up there. She would stop at Anchor Point on her way. I hadn't seen either place, and Austin was at the Point! Again, "I want to go, too."

"We were at Anchor Point... I had been so eager to get there, and was just as eager to leave a week later. There could hardly be a place which held more discomfort for a woman; but as the *Canby* swung in to harbor that fine morning, it seemed a haven of refuge.

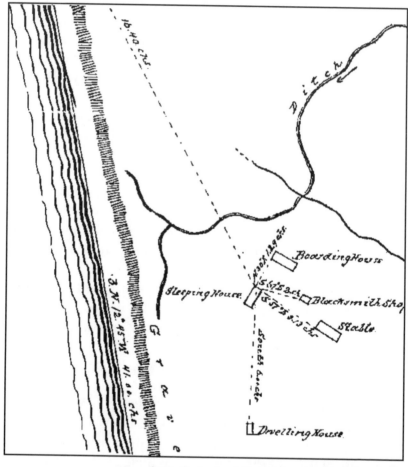

from Order for Mineral Survey No. 259, 1896

How many buildings shown on this schematic of the Anchor Point gold camp were constructed remains unknown. Pennock's crew worked there along with miners from allegedly competing companies.

The whole country around Anchor Point was a bog. Luckily I had worn rubber boots for the trip. Not far from the camp were some cabins in which the Indians lived. One day Fritz... asked if I would like to go over and see the Indians.

A half–Russian man with his wife and three children lived in one cabin. He spoke little English; she, none. But she smiled in a friendly fashion, and we all shook hands.

The one room was warm and clean, although it smelled fishy. The Indians caught fish and sold them to the cannery at Kasilof for one cent each. There wasn't much game in the country, what with so many white people milling around, and the Indians lived mostly on fish."

Five years before the Alaska Gold Mining Company crew arrived, gold–bearing beach sands were being worked at Anchor Point. Activities accelerated in 1896 when Homer Pennock arrived, supposedly with six hydraulic mining machines, 21 horses and 200,000 board feet of lumber. A lengthy ditch was dug to concentrate water so that hydraulic gear could wash the fine golden particles from the sands and gravels.

COOK INLET

Schooner Ella Johnson will leave Seattle April 12 for Sunrise and Hope City. Fare, $30, including 1000 pounds freight. Extra freight, $8 per ton. For further information apply to Capt. Smith, on board.

"The men had dug a ditch at the Point to bring water from a small lake for washing the beach sand and gravel. They tried out a sluice box as an experiment, and I heard they got about eighteen dollars in gold dust in two hours of work. Their greatest difficulty was the tide. Twice each twenty–four hours, at high tide, the water lapped the foot of the twenty–foot bluff, a rise and fall of twenty–five feet. After the sluice boxes had been washed away once or twice, they learned that twice daily the boxes had to be lifted up on the bluff and replaced as the tide ebbed.

The beach gold was very light and difficult to handle. They used quicksilver to amalgamate it.

Henry told Austin that he would like us to stay at Anchor Point through the winter as caretakers; but I told Austin that if he did, I'd

29

divorce him. It was a ghastly place on a boggy, half–cleared piece of land. The mosquitoes wore overcoats, and were half an inch long. They really were covered all over with a brown, furry coat.

Was this the glamour of a gold–rush? The men grew more surly and irritable day by day. Were these the things which were to duplicate the romance and gay endeavor of the days of '49? Men wet, dirty, unkempt, swearing at each other, at the weather, at the lack of comforts, at the Company, and finally at the very ideal that had brought them north; men at best of ordinary fibre, transplanted and being tried in a crude and severe school. Gold–hungry, eager to make their pile and get back to the pleasure the gold would purchase, the North was rapidly bringing out the best and the worst in them all."

from *A Game of Bluff, The Alaska Sportsman, October, 1945*

Doomed to Disappointment

1848	Peter Doroshin discovered gold near Cooper Landing
1888	Alexander King discovered another deposit on the Kenai Peninsula
1893	gold is found near Hope
1894-96	the Cook Inlet rush begins, falters, begins again

"In the fall of 1894 it was rumored that rich deposits of the precious metal [gold] had been discovered, and during the spring and summer of this year about 300 miners visited that locality for

the purpose of working the placers of which they had heard such glowing reports. But they were doomed to disappointment, and failed to find gold in paying quantities. After enduring much hardship and privation, they returned home and the placer mines of Cooks Inlet are deserted.

"Later in the fall, and after nearly all the miners had left this field of unprofitable labor and returned to their homes, it was reported that rich and extensive gold placers had been found in the vicinity of the claims that had so lately been abandoned. This report was given extensive circulation and caused great excitement among gold miners in all parts of the country. Early in the month of February last the tide of travel commenced to set in toward Cooks Inlet... By the 20th of May 2,000 men had arrived... and were encamped upon the supposed rich mining ground, which was at that date covered with 6 feet of snow. About the 20th of June the snow disappeared... Little or no gold being found, disappointment and misfortune was the lot of 95 per cent of all those who had gone to this unfortunate field.

"The steamship Excelsior left this port on September 7 for Cooks Inlet with the intention of bringing back to Sitka as many miners as the capacity of the vessel will allow. This steamer will make four voyages to the Cooks Inlet mines and return, which will enable a majority of the men to leave there before the coming of the winter months. Passengers arriving here connect with steamers for Seattle and other points on Puget Sound. There undoubtedly is gold in this part of the Territory. Large quantities of quartz rock have been found which will justify investigation. A few placer claims were discovered which paid for the working. Two incorporated companies are making expensive and permanent preparations for using hydraulic machinery on claims which they own, believing that gold can be found in paying quantities if these mines are properly manipulated."

from the Report of the Governor of Alaska, 1896

In 1896, the Spit served as a staging area for gold miners. Ocean-going vessels anchored just inside the tip of Coal Point and disgorged masses of men and mountains of gear to await smaller vessels to transport them up Cook Inlet. Seattle and Sitka newspapers regularly printed articles about Alaskan-bound adventurers.

John McArthur, a gold-seeker, wrote to a Seattle friend about his trip to the Kenai River. He mentioned what may have been the first tent city on the Spit.

May

There were nearly 300 men encamped at this point awaiting transportation to Turnagain arm, and great care was required to get the party off without causing a stampede to the Kenai river. Everything being favorable, the start was made on May 7. Early on that morning two heavily laden dory boats left Coal point bound up Cook Inlet. We camped that night near Anchor point..."

Later that summer, as McArthur's party was leaving Cooper Landing, they found this ditty written on a cigarette card:

'In God we trusted,
In Alaska we busted.
Let her rattle:
Will try again in old Seattle.'

"From Nenilchik we proceeded to Anchor point. Here the Boston and Alaska Mining company is working a hydraulic claim of several hundred acres, the only one on Cook Inlet. It has completed over five miles of ditch and employs about seventy-five men. The manager had just arrived on the steamer Gen. Canby, bringing a large amount of supplies, lumber and merchandise. Mr. Clark, an experienced hydraulic mining engineer from California, is in charge of the work. Mr. Lacy a deputy United States mineral surveyor, is in charge of that branch of the work.

"The Gen. Canby *was going to Coal point to get a supply of coal and Mr. Thorton invited us to take passage on her. Our boat and outfit were hoisted on deck, and after a pleasant run of a few hours down the inlet, we arrived at this place, after an absence of two months.*

"Coal point on Kachemak bay has one of the finest harbors on the whole coast. There are thousands of acres of fine coal at this place awaiting capital to develop the mines. When I left here last May this was a tented city of several hundred men, but at present it contains only some frame buildings and log houses. The bark Theobold, *of San Francisco, is here waiting for a crew to take her back, the sailors having deserted her and gone to the mines. The steam schooner* Perry, *of Puget sound, now running in Alaska waters, is taking coal."*

John McArthur
from *Alaska Searchlight, August 29, 1896*

About this time, Homer Pennock and his crew were in the news.

March 18, 1896

"Excelsior sailed with 60 passengers, six engines to be used for hydraulic mining, took 200,000 board feet of lumber on board at Tacoma. They also took 21 head of horses. The men were from Oakland and Eastern capital was represented.

"Dr. Walker and Homer Pennock are at the head of the expedition..."

from *Seattle Post Intelligencer*

May 4

"...there are large quantities of coal along the beach and much driftwood, so the residents of the camp will not suffer from the cold to any extreme. The Excelsior *people, who were the first to land, are well housed and have stores enough for several years at least. They are up in that country to make money... The* Lakma *men say*

ORDER FOR MINERAL SURVEY.

Department of the Interior,

OFFICE OF U. S. SURVEYOR GENERAL.

Sitka, Alaska
August 6, 1896

To A. Lacey
U. S. Deputy Mineral Surveyor.
Kodiak, Alaska

SIR:

Application has been filed in this office by the Boston and Alaska Company dated August 8, 1896, for an official survey of the mining claim of the Boston and Alaska Company known as the California Placer Mining Claim situate in the Anchor Point mining district. District of Alaska (County in Section _____, Township No. _____, Range No. _____) which claim is based upon a location made on May 14th 1896, and duly recorded on August 3d 1896 and is fully described in the duly certified copy of the record of the location certificate, filed by the applicant for said survey, a copy of which is herewith inclosed. You are hereby directed to make the survey of said claim in strict conformity with existing laws, official regulations, and instructions thereunder, and to make proper return to this office. Said survey will be designated as Survey No. 259

Very respectfully,

Louis L. Williams
Ex officio — U. S. Surveyor General for the Dist of Alaska

that the settlement presents a very picturesque appearance with its numerous tents and busy little streets."

from *Seattle Post Intelligencer*

May 19

"Among the Albion's *return passengers were... Pennock and Dr. Wilmers of Oakland who went on the* Excelsior *and got as far... as Coal Bay. Mr. Pennock says the party is now perfectly willing to stay there, owing to the remarkable finds of coal and gold. They were camped in a pleasant spot, entirely clear of winter, although only four miles away were glaciers and perpetual snow. Mr. Pennock says he has returned to purchase sufficient supplies for the party to remain at Coal Bay for a couple of years. He is exceedingly enthusiastic about the country.*

"If we haven't struck a patch that will put the famous Comstock into the shade," he said last night, "then I miss my reckoning. We have located a limitless number of prospects and if there is a square yard within its boundaries that will not yield a dollar, then I'll confess to the untruth and give the fellow who catches me $1000. The mountain has been figured out to pan $3.00 to the square yard... That the entire country is rich, there is not the slightest doubt, but a man going in there must take his chances and should not think of venturing without first providing himself with a first class outfit, together with sufficient ready money to maintain himself while there, and more important, to pay his fare home again should he not meet with good luck. In Cook Inlet Country, it is the same as any other business venture. Everyone cannot succeed, and every man should provide against possible failure."

from *Seattle Post Intelligencer*

June 6

"Members of the Bryant's *crew reiterated the statements made by others who have been at Coal Bay that there was a great deal of suffering among the people camped on the spit from exposure and cold."* from *Seattle Post Intelligencer*

June 9

"Dr. R. K. Dunn of this city (Oakland) has returned from Cook Inlet with anything but a flattering report of the outlook for the hundreds of prospectors scattered through the snowy wastes of that locality. Dr. Dunn declares that the truth had not been told concerning the gold fields. He left Oakland in February last with the Walker-Pennock expedition. He had returned very much disgusted and satisfied that gold mining in Alaska is an elusive dream."

from *Seattle Post Intelligencer*

September 15

An Oft Told Tale, More Discouraged Prospects from Cook Inlet

"They are working on top of a coal seam anyway and never in the history of mining has there been found a gold bearing country over a coal vein."

from *Seattle Post Intelligencer*

1896

"Anchor Point camp was a huddle of tents and low buildings of rough boards hastily thrown together. The endless bog and the overcast days added to the hopeless desolation of its appearance. Three times daily came the excitement of splashing through ankle–deep mud to the cookhouse.

On June first I came back to Homer, partly on horseback and partly on foot, since I dismounted to get a drink about halfway, and was unable to mount again. I led the horse along the beach, sometimes where there was barely room for us to walk between the cliff and the high tide.

drawing by JoAnne Heron from a photograph by Della Murray Banks
"When it wasn't raining, it was pleasant to explore the Homer spit. I often walked to the crescent–shaped grove of trees near the bluff, where the flowers were lovely."

Back at Homer, life went on much as before. I found some diversion in taking pictures, and although it was difficult to do any developing and printing, I did get some fairly good shots. By the middle of June there were many wild flowers in bloom – violets and white stars on the spit, wild currants in the groves. The days were so long there was really no night. The sun set after ten o–clock, like a big red ball behind Iliamna Volcano; the ice of Grewingk Glacier, across Kachemak Bay, grew pink soon after two in the morning, when the sun crept over the mountains. It was hard to tell the directions with the sun everywhere.

It was impossible to find out what success the different prospecting parties were having. All questions were received with a "why don't you mind your own business" attitude, which affronted my newspaper mind accustomed to the view that anything you wanted to know was your business.

The *Canby* came down the Inlet on June 16, to prepare for the trip to Kodiak. I was sick with disappointment when Austin wasn't aboard, but he walked down a couple of days later.

It was finally settled that Ben, a deep–sea sailor, Vail, the assayer, Young Ed Guilbault, and I as cook, would go to Kodiak."

Della Murray Banks
from *Prospecting Trip to Kodiak,*
The Alaska Sportsman, November, 1945

The rest of the article detailed another disappointment, no gold and a tragic accident. Della Banks badly burned two fingers.

"Back at Homer once more on July 31, everyone seemed glad to see us. Dr. Rogers dressed my hand immediately, and although he said it was in bad shape, he seemed to agree with the Wood Island doctor that my fingers might be saved.

It seemed the Company prospects were favorable. Pennock had been beaming when we saw him at Wood Island. The Eastern men were more than pleased with what they had seen at Anchor Point.

This wasn't a poor man's country, we told ourselves, nor a country where things could be done in a hurry. With all the equipment our Company had and the money they were putting in, our fortune would come. We had been in too much of a hurry. Our optimism revived somewhat.

Late August brought high tides, flooding the center of the saucer–shaped sand spit and making an island of the little grove of trees. The king salmon were running, and at each ebb tide they would be caught in the flooded hollow. The men got forty–eight once. Nice salmon, but I didn't care for any more fish. Enough is enough!

Odd, the things you remember and the things you forget. When I hear now of Kodiak Island, I seldom think of my fingers. What comes clearly to my mind is the cliff of blue forget–me–nots rising from the water, blue as the sky from water–line to skyline. It is the same with the Homer spit, the little island where the wild currants grew, the tiny blue violets, and the pink and white stars. Such things never disappoint you as a dream of fortune does.

A three–day cleanup of the hydraulic giant at Anchor Point yielded between three and four thousand dollars in gold; at least, that's what they told us. We didn't see it.

Lane, an Eastern expert who had been at the Point since July, told us the Company prospects there were entirely satisfactory, and that we had a splendid thing. Ah, we thought, that fortune may not be far away, after all!

Nature did her best to liven things up for our anxious, weary group. On September 7, Iliamna Volcano belched out a great volume of smoke. On the 17th we had our ninth and worst earthquake, and Chinnaburo (Augustine Volcano) smoked a little. Evidently she was not dead, after all.

The *Excelsior*, joy to her name, was remodeled for passenger service and put on the run between Sitka and the Inlet.

On September 18, we left for Ninilchik in the converted whaleboat... Ninilchik, a village of about forty Indians and Russians, straggled up the hill from the triangular flat at the mouth of the creek. We pitched our eight–by–ten tent, and set up our crowded camp.

Here, at last, I met Mrs. Cooper, the Russian wife of the kind gentleman who had visited us at the spit in the spring. She was beautiful.

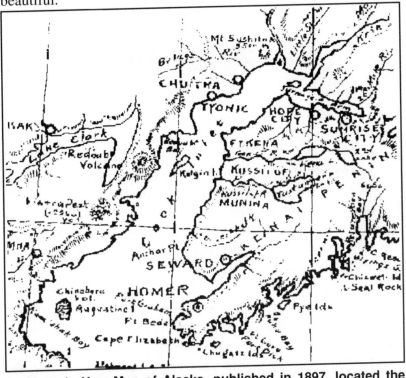

Anderson's New Map of Alaska, published in 1897, located the community of Seward on the north shore of Kachemak Bay, Homer on the south. While the placement of Homer was a mistake, a post office of Seward did exist briefly near the coal camps at McNeil, Cottonwood and Eastland Creeks.

At last, on September 30, we were aboard the *Dora* on our way home! All the crowd from Homer were with us except Von Hasslocher and Hughes, who stayed as caretakers, and Cosad and Mr. Hand, who went on the Marion.

Passing Mary Island, the old Customs port, we were out of Alaska. It was almost seven months since we were anchored there before. I had seen a lot of country, but just then I'd have preferred to see a nice tub of hot water rather than Muir Glacier, which I thought was about the only thing I'd missed.

How wrong I was! The lure of that golden fortune, always at the end of one more summer's work, was to take me back to Alaska again and again."

from The Alaska Sportsman, December, 1945

1897

Like many people far from home and familiar surroundings, the coal miners in upper Kachemak Bay wanted mail. In July 1895, Freeman Curtiss applied to be postmaster of a new office to be called Seward. Located near "Upper Cottonwood" creek, Seward was projected to serve 100 residents. Apparently the name was accepted yet within the year, "Seward" was eliminated and re-placed with Homer. The physical location of the first Homer post office remained near McNeil Canyon and on October 3, Stephen T. Penberthy was appointed postmaster. In 1897, the office moved to the Homer Spit.

The annual salary of the postmaster was determined by the volume of business his office transacted. In 1898, Penberthy's salary was $25.00. Postmasters in Kenai, Seldovia and Tynonok received no salary in contrast to those in Juneau, who earned $1,800 a year, and in Skagway, $1,600!

Postmaster Penberthy, of Homer, provided the underlined information to accompany the 1897 postal route map. In 1907 the Homer post office closed and mail was routed through Seldovia. The town of Homer had been abandoned since 1902 and only a company watchman lived on the Spit.

Post Office Department,
Topographer's Office,
Washington, D.C., <u>March 30, 1897</u>

To Postmaster at <u>Homer/Late Seward</u>

The (P.O. Dept.) name of my Office is <u>Homer</u>
The name of the most prominent river near it is <u>Between the Copper and Yukon</u>
The name of the nearest creek is <u>Kassiloff</u>
The name of the nearest Office on Route No._____ is <u>Tyoonic</u>, and its distance is <u>100</u> miles, by the traveled road, in a <u>NW &W</u> direction from this, my Office.
The name of the nearest Office, on the same route, on the other side, is <u>Kodiak</u> and its distance by the most direct road is _____ miles in a direction from this, my Office.

(Signature of Postmaster,) <u>S. .... Penberthy</u>
(Date,) <u>June 30/97</u>

"Lack of money in Alaska and lack of money in California were two entirely different things. We did not go back to Denver–Austin never saw Denver again, nor did I for twenty–five years but stayed in California pinching pennies, a hardship I had not had to face along with the other difficulties of the previous summer.

Still, there was always our stock in the Alaska Gold Mining Company, Austin's compensation for his summer's work, which would make us rich at some early day–we hoped! Pinching pennies wasn't so hard with wealth just around the corner; and it was unreasonable of us to have expected a fortune in one summer. Conditions at Cook Inlet had been more difficult than anyone had anticipated. Next year would be different!

Next year came rapidly. We spent all of our spare time...talking of the coming year in Alaska. Hard times in the States made Alaska seem all the more a land of opportunity. The great adventure was still an adventure; our belief in the future fed upon hopes and wishes, and grew stronger day by day.

We left San Francisco on April 8, 1897...

We had heard that the big house on the Homer spit had burned during the winter. It was true! We reached Homer at noon of April 27, and the big house was gone. The Company had left large stores of supplies in the house, but not knowing how much had been saved, everything had been duplicated at Sitka. Without the house, there was little left of Homer!

It seems that the day after Thanksgiving, a heavy snowstorm struck the spit. In the darkness of early morning, the fire broke out. Nine people were living in the building–a family whom

I shall call Sills, man and wife with two small boys; Hughes and Von Hasslocher, the caretakers; and three other men. The log building was dry as tinder, and no means of fighting the fire were at hand. Nine shivering human beings stood out in the snow storm, seeing most of their provisions and clothing go up in smoke.

At the risk of their lives, they saved what they could. Then they huddled together in the best of the shacks with the little they had until the storm subsided, and someone could go to McNiell's coal camp for help. Someone had pulled the head of Mrs. Sills' sewing machine out the window. The metal frame was salvaged after the fire burned out, and she had saved three needles. Von fitted a wooden table to the frame of the machine, so Mrs. Sills at least had something to make clothes on.

When the men brought the blankets from NcNeill's, she made underclothes, socks, a dress for herself, everything for nine people to wear through the winter, out of blankets!

Being low on provisions, they put themselves on rations. Even then, they ran out of sugar, meat, and butter by March. They built a house of sorts out of what had been a barn; and in it they existed on that desolate sand spit, while their food supplies dwindled... Mrs. Sills declared she mourned most the more than ninety quarts of currant wine and sixty glasses of currant jelly they had lost!"

* * *

"It really seemed good to be back at the sand spit again. The company built a new log house on the site of the old one, not so large as the former one but with a tighter foundation. The high tides of November had brought six inches of water into the old house. Moving into the new house took place while we were there.

I busied myself making two sails, and tarpaulins to cover the cabin and deck of the new sailboat.

We heard here that there had been trouble again at Anchor Point. Two of the Eastern men, a number of the workmen, and Heney and Rice, the ditch contractors, had left on the *Dora*. At Sitka the men demanded that the Company pay their fare to Seattle; and in the argument both Heney and Rice had been shot, one in the arm and the other in the breast. The Eastern men hid in the hold of the *Topeka* until she sailed. Heney, by the way, was the Michael J. Heney who became noted later for his remarkable work on the White Pass & Yukon, and the Copper River & Northwestern Railroads."

On a visit to Tyonic and Ladd's Station, "...Austin told me of a man who wanted pictures of the work being done on claims there,

from the Anchorage Museum of History and Art, B90.3.13

Built in 1880 for the Alaska Commercial Company, the *Dora* served Southeast, Southcentral and Southwest Alaska as a mail steamer and passenger ship.

to send to his company in the States. Next day Austin took me up in the rowboat. I got ten dollars just for going, and an order for six dollar's worth of prints of the snapshots I took, all paid in advance. Next day we went back with the finished pictures, and he ordered another five dollar's worth. My entire outfit had cost only twenty-five dollars. My extravagance turned out to be the only money-maker among us, paying for itself and far more, besides. Everyone wanted pictures, and I was the only one equipped to supply them. Now, nearly half a century later, those pictures I took, developed, and printed with an outfit that would seem hopelessly crude to the modern photographer, have brought me pleasure beyond all monetary consideration.

In later years when we met again, the men of the Company white–haired and strangely modified by time, we recalled the "good old days" over my snapshots. Now that they are all dead but Von Hasslocher and me, those pictures invoke the dreams that help me through the long, weary, lonely hours and days and years. Oh, to be there again–with Austin–on the Cook Inlet of 1897!"

In August, 1896, George Carmack, Tagish Charlie and Skookum Jim discovered gold on Bonanza Creek and, like the ebbing tides of Cook Inlet, the miners surged out of Southcentral Alaska into Canada. Della and Austin Banks were in Sunrise City when the strike was announced. She wrote,

"Then the *Perry* came on August 15, bringing news that electrified the camp and drained it of most of its inhabitants.

The story of the Klondike stampede has been told over and over again, sometimes by men who lived it and know whereof they spoke; often by those who only heard of it, or had vivid imaginations. I can only tell how it affected the little settlement of Sunrise...

I shall never forget that night: Captain Austin Lathrop stood on the steps of the store and told the growing crowd of the ship–load of gold from the Klondike–that "ton of gold" which sent thousands stampeding into the little–known Yukon.

"I'm going!" was the universal cry. The *Perry* and the *Stella Erland* left on the flood tide, and on them left men who had simply dropped everything and stepped aboard. Within twelve hours, Sunrise was practically depopulated.

Our boys returned, and cursed their luck at having missed the excitement. Austin... could talk of nothing else. After we had gone to bed, he turned on his blankets toward Ed.

"Well, Ed, shall we go to the Klondike?"

"I have to see my children this winter, and then I'm ready to go," Ed answered.

"When do we start?" I cried, sitting up quickly.

Ed glanced at me. "You have got pluck," he commented. But if Austin went, of course I would go.

We left Sunrise on the high tide of August 17, reached Homer early in the morning of the twenty–fourth, and sailed that night on the *Dora* with the remainder of the sand–spit crowd. Everyone was bound for the Klondike; some directly, some after a brief visit home."

Della Murray Banks,
from *Hope Springs Eternal;*
The Alaska Sportsman, January, 1946

1898

With the migration of gold miners from Cook Inlet to the Klondike, even fewer people visited Homer. The post office and the opportunity to purchase coal kept the tiny community on the Spit alive.

Stephen Penberthy, standing , postmaster and caretaker of Homer from 1896 until about 1907, was often the only permanent resident after Homer was abandoned in 1902. Although Homer was never a Native village, Native residents such as Nicoli from Seldovia, Port Graham or English Bay (Nanwalek) were hired as day laborers.

courtesy of the U.S. Geological Survey

"Homer, which is a postoffice station, situated sixteen miles above Saldovia, is the headquarters of a mining company. It has a store and warehouses and a permanent population of about six people."

from an 1898 Alaska Commercial Company brochure

When the public demanded more information about the Klondike gold fields and Dawson City which sprang up overnight at the confluence of the Klondike and Yukon Rivers, the United States government ordered several military expeditions to Alaska to find routes to the Yukon River. The expedition headed by Captain Edwin F. Glenn passed through Kachemak Bay en route to upper Cook Inlet and the Interior. Glenn visited Homer.

"October 9.–...I took passage on the Dora *as far as Homer. Here I would meet the mail steamer of the Pacific Steam Whaling Company. The vessel at that time did not come farther up the inlet than Homer.*

"October 10.–I arrived at Homer at 3.30 a.m. and went on board the Perry, *which was lying at anchor in the bay. In the morning I went ashore and found that Homer consisted of a sand spit running out into the sea for a mile or two, thus forming a very good and well–protected harbor.*

On the spit were located seven houses and a tent. One of the houses is used by the company as a sort of hotel. There seems to be no water or wood on the spit, but plenty of coal in close vicinity, at Coal Bay. Although a post office had been located at this point during the past summer, it is not a suitable place, for the reason that during the high spring tides the major portion, if not all, of the spit is covered to a depth of a foot or more with salt water.

After examining the place I again went aboard the Perry. *We steamed up into the sea for a short distance to what is known as "Coal Banks," where the steamer expected to take on a load of coal as soon as the tide ran out. The banks are located about 10 or 12 miles from Homer... The coal is bright, clean to handle, and is readily broken off, with picks, in large pieces. The officers of the* Perry *said they had used it for the past three years and that it burns*

48

up completely... This was destined for Sunrise City, to be used for general purposes by the inhabitants during the ensuing winter."

Pennock and many of his crew, such as Della and Austin Banks, rushed to Canada in 1898. Unlike thousands who crossed the steep, treacherous Chikoot Pass, they followed the lesser traveled, less difficult, yet longer Chilkat Trail to connect with the Dalton Trail.

Between Pennock's departure from Homer in 1897 and his arrival in Canada, his leadership role changed. Now, he was associated with a group of thirty–six men from 10 states, England and British Columbia. Led by a former U.S. Cavalry man, Lieutenant Adair, the furtive group drew the attention of William D. MacBride, another gold–seeker who wrote in "The Story of the Dalton Trail,"

"As they were very reticent in their remarks and no one was able to discover their real business, they were called "The Mysterious 36" and later on in the season dubbed "The Big Push."

"It was understood, but not officially, that the Lieutenant was representing the Standard Oil Co. and some eastern Canadian capitalists."

photograph by H. D. Banks

Pennock's Post, which consisted of one log cabin, provided refuge for gold–seekers on the Dalton Trail. Because the eighteen by twenty foot cabin was built, in part, by inexperienced men, the leaky roof was covered by tents to help shed rain.

from Yukon Archives/MacBride Museum Collection

"In September, 1898, the party of 36 returned to Seattle with a pitifully small amount of gold and memories of a beautiful but harsh country whose hidden treasure was protected by swamp, quicksand, flash floods, treacherous rivers, deadfall, snow, ice, cold, mosquitoes and black flies."

Another Grand Adventure

Austin Banks went back to the Klondike in 1899 and, once more, Della went North to join him. Aboard the ship to Skagway, she encountered Homer Pennock.

"Finally in the spring of 1898 on learning that we would get no outfit at all, he (Austin) came to Seattle where he made arrangements to go with a party of nine men from Massachusetts and Connecticut on a trip over the Dalton Trail to a place where they expected to find gold and if not they would go on to the Klondike. He would get $50 a month and the trip would be two months. After I joined him in Seattle it was agreed that I would cook for the party at $50 a month on the trail and if on reaching the Yukon River I wanted to go home, then Dennison Tucker, the head of the party, would pay my expenses out to the coast and see me safely over the Pass to Skagway which he did...

But now on this trip to Dawson in 1899 I was strictly on my own...

Four times I had seen Seattle fade into a curtain of rain or fog behind me and set my face to the North in search of that illusive will–of–the–wisp...GOLD! Four times filled with hope and faith (with Austin the faith never failed, but with me while hope persisted, faith surely and steadily faded.) Four times I had sailed down the beautiful Inside Passage tired, disappointed, discouraged, ready to call it a day and settle down to civilized life in the States and yet each time as we neared port and saw the lights of Seattle... there was a sense of let–down, of anti–climax. Surely we were tired of hardship and wanted the comforts we had been so long without; safety, security... With all the disappointments each year had been a Great Adventure. Were there to be no more adventures?

Looking back to those days, a half–century and more in the past, it seems as if we turned North in the spring as regularly as the wild geese made their flight and with less real purpose. The geese at least knew where and why they were going while we only knew that we were going. A few years ago..., I told an old Alaskan friend that I meant to write a book of our life in the North. He asked the title I

photograph by Della Murray Banks

"Although I was a "mere woman," I'd had more experience on the trail than the men in our party. Austin and I were seeking our fortunes for the fourth time. The other men hadn't known what to expect when we set out for the North. During a pleasant summer's outing, they would pick up handfuls of gold, then go back rich in the fall. No cold winters and backbreaking work for them—so they thought."

51

had chosen. "As Geese Go North," I said smiling. He repeated it slowly twice, then an understanding grin broke on his face, "Yes, Della, and they still do!"

Among the *Humboldt* passengers was Homer Pennock, manager of the mining company of our Cook Inlet days. I had heard... that he was in Seattle and going north but had hoped he wouldn't be on the same boat for our unfortunate financial experience with him and his company was too recent. Austin had know him for years and trusted him too much and each time had been left "holding the bag". The boat was small, there was no way to avoid meeting him. The year before in Skagway I had been interviewed by the Skagway paper on our trip over the Dalton Trail to Shorty Creek and the fiasco it had proved to be not only for us but also for a party Pennock had sent in some months before us. I had told the editor the truth about the place and someone had sent Pennock a copy of the paper. Soon after... I had received a letter from him in which he had objected to any further interviews and had implied that he wanted me to hold my tongue. Naturally I had resented this... his letter made it an order... but I had not answered his letter. Now, on meeting him again, I commented on his letter saying I could not understand his writing to me in that way. "Because what you might say could interfere with my plans, because I wanted you to keep still."

"Oh, yes, I understood that part... that you wanted me to keep my mouth shut, but why bother to write and tell me that?"

Pennock said, "Yes, I know," in a disgusted tone. Then he asked how I was fixed about getting to Dawson, if I had money enough. I felt that incipient paresis must have set in (for one or the other of us) and almost said, "Oh yes," in a cheerful tone but suddenly remembered a Skagway experience in '98 when I had almost gone broke and thinking of the money due my husband, I caught myself in time and said with emphasis, "Well, I can just get through."

Pennock said he was not rich just then but he could let me have $20 if that would help. It would so I was that much ahead especially as we had given up hope of ever getting anything. Pennock said that if he ever got on his feet again, he would make it up to Austin, but he had made so many worthless promises... and $20 is $20.

Perhaps our Cook Inlet crowd were especially unlucky for not one of us ever "struck it rich".

from *"The Yukon Trip in 1899"*
unpublished by Della Murray Banks

1899

Homer stagnated after the first rush of gold–seekers and coal miners. Primarily, it was the presence of the post office which attracted people to the forlorn settlement. Among the ships anchoring in Kachemak Bay to post letters was the *George W. Elder.*

The Harriman Alaska Expedition

One of the last, major, privately–funded scientific expeditions of the nineteenth century occurred when Edward H. Harriman, an East Coast railroad magnate, assembled an impressive roster of the country's leading authors, artists, and scientists to join him and his family on a three month junket along the coast of Alaska and Siberia. Aboard the *George W. Elder* were such notables as John Muir, John Burroughs, G. K. Gilbert, Henry Gannett, C. Hart Merriam, William H. Dall, Louis Agassiz Fuertes, Charles Palache, Frederick Dellenbaugh, Benjamin Emerson and George Bird Grinnell. Others, such as Edward Curtis who later became a prominent photographer of Native American people, were just starting their careers. An impressive array of information and illustrative material was amassed by the individuals and eventually published.

A few paragraphs in the Report of the Governor of Alaska for 1899, noted that,

courtesy of the Smithsonian Archives, 88–16754

The *"Good Ship George W. Elder"* cruised about 9,000 miles along much of the coast of Alaska and Siberia from June through August. On board were about 28 prominent or soon–to–become prominent Americans documenting information about the features and re-sources of Alaska.

"It is seldom that so many men eminent in science and literature ever were afloat together upon one vessel for a cruise. All matters of detail ... were given personal attention by Mr. Harriman. The results of this expedition can not but be beneficial to Alaska. All these well–known scientists will have much to say upon their special subjects... The incredulous will take their testimony as coming from unpredicted sources so far as the physical features of Alaska are concerned. We in Alaska hope that this will be the forerunner of many similar expeditions."

Charles Palache, a mineralogist at Harvard University was aboard the *Elder*. He, William H. Dall, and G. K. Gilbert visited Halibut Cove and hiked to Grewingk Glacier which Dall had visited years earlier. Palache, later a prominent crystallographer and mineralogist, helped develop the Mineralogical Museum at Harvard University. His papers, related to the Harriman Expedition, offer brief glimpses of Homer.

Sept. 9th, 1975

"When Prof. Emerson invited father — then an instructor at Harvard — to go as his assistant on Mr. Harriman's Alaska

Expedition, father said "No, he couldn't go, he was getting married." Prof. Emerson replied, "You can get married any time, but there is only one Harriman Expedition." So father went and the letters he wrote to mother were all the journal that he kept. They were put in a file and no one knew of their existence until I began assembling records from father's files for the Harvard archives last winter. Mary Palache Gregory

Homer
June 30th, 1899

"I am writing now on Friday morning June 30th while lying at anchor off the tiny town of Homer in Cook Inlet, our next post office. We reached here this A.M. in lovely sunshine and are waiting to go ashore as soon as plans for our stay in this vicinity are complete. Homer is a lovely [lonely?] spot on the end of a long sand spit that juts out into Kachemak Bay. The surroundings however are attractive – low wooded mountains on one side and fine mountains with big glaciers on the other."

Kodiak
July 4th

"I sent my last letter at Homer. We expected to stay some time in Cook Inlet but for reasons known only to Mr. Harriman but chiefly having to do with the hunting we turned sharp around the same day and ran over here."

Saldovia
July 21st

"Land a party at Saldovia, Cooks Inlet at 5 A.M. I am up expecting to be put ashore at Homer soon after but delay follows delay and it is not till 10 A.M. that Gilbert, Dall, Coville and I get off in the launch. The early morning view of the great volcanic peaks, Uiamna and Redoubt on north side of Cook Inlet. We expect to spend two days near Homer while ship goes up the Inlet. Spend the remainder of the day at Halibut Cove, south shore of Kachemak Bay with interesting geology but fearfully bad travelling in the forest. On the way back to Homer for dinner we meet to our surprise

55

John Burroughs, wrote in *Far and Near*, that on June 30 the *George Elder* "...dropped anchor behind a low sandspit in Kachemak Bay, on the end of which is a group of four or five buildings making up the hamlet of Homer. There was nothing Homeric in the look of the

the Elder *turned back from her trip and get aboard for dinner. Pick up the other party about midnight and make straight away for Yakutat."*

As the *Elder* steamed out of Kachemak Bay, it left behind an almost treeless sandspit, a few derelict buildings, a post office and a postmaster. Yet, as the century ended, Homer was on the verge of expansion.

Months later, another ship anchored in the lee of the Spit. Offloaded were the accoutrements for a new town – cut lumber, windows, doors, metal roofing, rails for a railroad, tools, livestock and farming equipment. Hastily, before the winter winds swept down Kachemak Bay, employees of the Cook Inlet Coal Fields Company built the first town of Homer.

The tiny town was owned, occupied and managed by the Cook Inlet Coal Fields Company of West Virginia which was organized, in part, to acquire and develop coal lands, to construct railroads and to acquire town sites.

Capital stock in the Coal Fields Company was valued at one

place, but grandeur looked down upon it from the mountains around, especially from the great volcanic peaks, Iliamna and Redoubt, sixty miles across the inlet to the west."

million six hundred and fifty thousand dollars, was divided into sixteen thousand five hundred shares valued at one hundred dollars each. Of the 180 stockholders listed on the 1902 revised Certificate for the Cook Inlet Coal Fields Company, the majority were residents of Pennsylvania and New York. Among them were names familiar to the Homer of 1896–97: Guilbault, Penberthy, Ray and Pennock.

The Bourse was built to house exchanges and groups promoting the commercial interests of Philadelphia. The Cook Inlet Coal Fields Company, with its many Pennsylvania stockholders, headquartered in The Bourse at the turn of the century.

courtesy of The Bourse, Philadelphia

Epilog

Della Murray Banks

1864   born Della Murray..., December 3, Rochester, Minnesota

1893   married Austin Banks, Denver, Colorado

1894–96   Denver City resident, proofreader at the Denver Times

1896   summer, headquartered on the Homer Spit

1897   summer, headquartered on the Homer Spit wintered on Prince of Wales Island, Southeast Alaska

1898   summer, gold–seeking on the Dalton Trail, Yukon Territory

1899   gold–seeking in the Klondike

1900   gave birth to Sydney Allen Banks, in California

1901   traveled to Skagway, with baby, to meet Austin

about 1901–1919   family lived in Seattle

1919   Austin died in Seattle

about 1919   Della and Sydney moved to Los Angeles County, California

1921   Della visited Denver after 25 years absence

1923   fell from a horse, confined to a wheelchair

1950   died in Los Angeles, California, age 85

"Years ago, Austin took the trail into the unknown without me; but I shall follow soon. ...through the Alaska newspapers I follow the events of the North. The names of many old friends are there—mostly in the obituary columns these days. They'll all be there in that prospector's heaven, where we'll go searching for the pot of gold at the rainbow's end."

Della Murray Banks, from *The Alaska Sportsman, February, 1945*

Austin Banks

1851	born in Indiana
1879	lived in Colorado where he appeared to be very active in the Denver fire department and on local running teams
1893	married Della Murray in Denver
1881–96	Denver City resident
1896	summer, headquartered on the Homer Spit
1897	summer, headquartered on the Homer Spit wintered on Prince of Wales Island, Southeast Alaska
1898	summer, gold–seeking on the Dalton Trail, Yukon Territory
1899	gold–seeking in the Klondike
1919	died at age 67 in Seattle, Washington

from the collection of Roger Banks

Weary from five seasons of gold hunting, Austin Banks and his family retired to Seattle where he worked as a shoe repairman until his death in 1919.

Homer Pennock

In 1900, after failing to find his fortune in the Klondike, Pennock returned to California where several crew members from the Homer expedition lived. The 1900 Census showed him and his wife, Lilian, residing in Oakland. According to it, she was born in 1860 making her 20 years younger than Pennock, yet in the column listing their ages, he was 60 and she was 29!

Homer and Lilian Pennock retired to New York sometime during or after 1900. The New York Times ran the following article that year. Pennock was listed in the city directory, but not in the social directory, from 1908 until his death four years later.

Homer Pennock appeared to have led a full life. He traveled extensively, created legitimate and illegitimate companies and worked with all types of people. From prison to prosperity to poverty, he pursued his passion for mining, his passion for money.

Homer Pennock Bankrupt

Mining Speculator's Schedules Show
$317,404 and No Assets

Homer Pennock, who is in the mining business at 52 Broadway and resided at the Hotel Earlington, has filed a petition in bankruptcy with liabilities of $317,404 and no assets. Of the liabilities, $242,167 are unsecured, $73,737 secured, and $1,500 accommodation paper. The debts were contracted from 1882 to 1898.

Among the creditors mentioned in the schedules are F.Q. Barstow of this city, $70,000 secured by 566 shares of stock of the Cook Inlet Coal Fields Company, par value $56,666; estate of N.A. Cowdrey, New York, $150,000 contracted in 1882; George Roberts, New York, $6,000 on a note given for purchase of an interest in a telegraph company; estate of Emmett Dinsmore, Brooklyn, $15,000 on a note; L.Z. Letter, Chicago, $25,000 money borrowed in 1882; John Brown, Chicago, $10,000 borrowed money; William Pemberthey, Chicago, $10,000 borrowed money.

Mr. Pennock has been interested in mining enterprises during the past twenty years in Colorado, and lately in Alaska.

from *New York Times,*
October 17, 1900

61

Homer Pennock, Once Millionaire, Dies in Poverty

Gained, With Associates, $5,000,000 From Operation of Leadville Mine

Boulder, Colo. July 27.–The death of Homer Pennock, a former Coloradan, at New York city, this week, at the age of 72, recalls to mind the gold fever in Colorado in the early days.

Although Pennock was once the owner of the famous Robert E. Lee mine at Leadville which netted him and his associates $5,000,000. and a man for years accustomed to deal in millions, he died practically in poverty.

Pennock came to Colorado in the early '70's on his way to the coast. After he had enough of mining in the Pacific states, he returned to Colorado and went to Leadville, where he opened the Robert E. Lee mine. It is recorded in the mining history of this state that $117,000 worth of gold was taken from this mine in a single night.

From Colorado Pennock went to Detroit, Mich., where he invested his millions in a corset factory. His next venture was a car wheel works at Pennock, Ill., near Chicago, where he owned 1,000 acres of land. He was worth about $4,000,000 when he tired of this activity and went down into the Transvaal to dig diamonds. It was there that he lost the greater part of his fortune.

Of recent years he had been living with his young wife in New York city and promoting mines in Colorado and Alaska.

from Denver Post

Although he left no children, certainly his name will be remembered – on a townsite in Ontario, a subdivision in Chicago, an outpost on the Dalton Trail and on this community in Alaska.

This, then, is a glimpse of Homer Pennock for whom our town was named, of Austin Banks who's faith in Pennock brought him to Kachemak Bay, of Della Murray Banks, Homer's first historian, and of the Homer Spit which was wilder and more wonderful a century ago. It was a time when all things seemed possible, when golden fortunes were awaiting in the next creek around the bluff.

When the citizens of Homer were considering incorporating as a first class city in March 1964, an editorial in the Cook Inlet Courier encouraged them to rename the community "...so the incorporated area could get off to a first–class start and not be encumbered by the name of the first con–man to practice on Coal Spit."

They chose to support Homer.

Homer Pennock

1840	January, born in New York
1871	jailed in New York City for his role in the Otter Head Tin Swindle; a townsite in Ontario, Canada was named Homer
1879–81?	strikes it rich in the silver mines of Leadville, Colorado
?	owned a corset factory in Detroit, Michigan
1883–85	president of an elastic steel car wheel factory in Chicago, Illinois; owns Pennock Village
?	to the Transvaal supposedly to dig diamonds yet it may have been gold
1890 ?	married Lilian G. ?, a Michigan woman. They had no children.
1896–97	headquartered on the Spit, searched for gold throughout Southcentral Alaska; left his name on Homer, Alaska
1898–99	gold–seeking in the Klondike; left his name on Pennock's Post on the Dalton Trail
1912	July, died in Manhattan, New York; age 73

Pennock's occupation listed on his death certificate was engineer. Although the cause of his death was uncertain, the diagnosis, two days before his death, was thrombosis of the posterior cerebral artery.

Select Bibliography

1872 Richmond, J. F.
New York and its Institutions, 1609-1872.
New York: E. B. Treat, San Francisco: A. L. Bancroft & Co .

1880 Kent, L. A.
Leadville. Denver: Denver Daily Times Steam Printing House
& Blank Book Manufactory.

1883 Department of the Interior
U.S. National Museum Proceedings of the United States
National Museum, Vol. V, 1882. Washington: Government
Printing Office.

1889 Andreas, A. T.
History of Cook County, Illinois. Chicago: A. T. Andreas.

Petroff, Ivan
Report on the Population, Industries, and Resources of Alaska,
10th Census: 1880. U.S. Department of the Interior, Washington.

1896 Alaska Searchlight, A Literary and News Journal of the Far
North. Vol. II, #38, Aug. 29. Juneau: E.O. Sylvester.

Lamb, Martha and Mrs. Burton Harrison
History of the City of New York: Its Origin, Rise, and Progress.
New York: The A. S. Barnes Company.

Dall, William Healy
Report on Coal and Lignite of Alaska in U.S. Geological
Survey Seventeenth Annual Report, 1895-96, Part I. Wash-
ington: Government Printing Office.

U.S. Surveyor General for the District of Alaska Order for
Mineral Survey No. 259, August 6, 1896.

1897   United States Post Office Department Questionnaire regarding location of post office.

1898   Alaska Commercial Company
"To The Klondike Gold Fields and Other Points of Interest in Alaska," brochure.

1899   Palache, Charles
Harriman Alaska Expedition, May-July, 1899. Letters and Journal. Unpublished.

1900   U.S. Census: 1900. State of California, Town of Oakland.

1901   Kelly, James
Plat of Water Terminus - Cook Inlet Coal Fields Co's. Ry.

1904   Burroughs, John
Far and Near. Cambridge: Houghton, Mifflin and Company.

1906   Stone, R. W.
Coal Fields of the Kachemak Bay Region in Mineral Resources of Kenai Peninsula, Alaska. United States Geological Survey. Washington: Government Printing Office.

1912   Standard Certificate of Death, Homer Pennock. State of New York. New York: Bureau of Records.

McKellar, Peter
"The Otter Head Tin Swindle" in The Thunder Bay Historical Society Fourth Annual Report, papers of 1912-13. Ontario.

1922   Cotton, Bruce
An Adventure in Alaska. Baltimore: The Sun Printing Office.

1939   Thompson, Arthur R.
Gold-Seeking on the Dalton Trail. Boston: Little, Brown, and Company.

1945   Banks, Della Murray
"Klondike Gold Fever", January - February 1945. Ketchikan: The Alaska Sportsman.

"Homer's Gold Seekers," October 1945 - January 1946. Ketchikan: The Alaska Sportsman.

1965 Ricks, Melvin B.
Directory of Alaska Postoffices and Postmasters. Ketchikan: Tongass Publishing Company.

1973 Barry, Mary J.
A History of Mining on the Kenai Peninsula. Anchorage: Alaska Northwest Publishing Company.

1981 Klein, Janet
A History of Kachemak Bay, the Country, the Communities. Homer, Alaska: Homer Society of Natural History.

Pierce, Richard A., trans. and ed.
Atlas of the Northwest Coasts of America. Kingston, Ontario: The Limestone Press.

1990 Pierce, Richard A.
Russian America: A Biographical Dictionary. Kingston, Ontario and Fairbanks, Alaska: The Limestone Press.

1991 Kalifornsky, Peter
A Dena'ina Legacy, K'tl'egh'i Sukdu, The Collected Writings of Peter Kalifornsky. Fairbanks: University of Alaska, Alaska Native Language Center.

1992 Tower, Elizabeth A.
Big Mike Heney, Irish Prince of the Iron Rails. Anchorage: Elizabeth A. Tower.

1995 Gregory, Judith Palache. Letter to author, August 3.

Newspapers

Seattle Post Intelligencer, 1896
The Alaskan, Sitka. August - October 1896
Yukon Archives Newspaper Clipping, MacBride, William D. "The Dalton Trail and The Mysterious 36". no date

Index

This very select index primarily lists names, places and events that were most frequently mentioned. Earlier spellings of place names are often listed in parentheses after the contemporary spelling. Bold numbers refer to an illustration or a caption. If related information is found in the text on the same page, that page number is not written in bold.

Miltimore: Elastic Steel Car Wheel Company 18; John 18

New York: City 15, 61, 62; state 14, 57; Stock Exchange 17-18; Times 61
Ninilchik **26**, 27, 32, 39

Otter Head Tin Swindle 14-15, 63

Palache, Charles 53-56
Penberthy, Stephen T. 40-41, **47**, 57
Pennock, Homer 14-29, 33, 35-36, 38, 49, 50, 52, 57, 61-63; Lilian 61, 63
Pennock's Post, Dalton Trail **49**, 63
Pennock Village, Illinois 18, 62-63
Port Graham 8, **47**; coal mine 6

Report of the Governor of Alaska 30-31, 53-54
Robert E. Lee mine 17-18, 62
Russian navigators 4-6

Seattle: Post Intelligencer 33, 35, 36; Washington 19, 23, 32, 44, 50, 51, 58, **60**
Seldovia (Saldovia) 8, 40, **47**, 48, 55
Seward (in Kachemak Bay) **39**, 40-41
Sitka, Alaska 32, **34**, 39, 44
Skagway, Alaska 27, 40, 50
Spruce, Sitka 1, 7, 12
Station Coal Point **10**, **11**. *See also* Coal Point
Sunrise (Sunrise City), Alaska 46, 49

Tebenkov, Mikhail 4, **5**
The Bourse **57**
The Mysterious 36 **16**, 49
The Tombs 15
Tyonic (Tynonok, Tyoonic), Alaska 40-41, 44-45

U.S. Survey No. 105 12

Yukon: schooner 6; River **41**, 48, 50

Homer Spit, 1996 photograph by Michael R. Dudash, Homer News

Other Books About Homer and Kachemak Bay

ARCHAEOLOGY OF KACHEMAK BAY, ALASKA
by Janet R. Klein

Written for the public and archaeologists-at-heart, this very readable, well-illustrated soft cover book describes the resources of Kachemak Bay, introduces basic field archaeology terms and methods, presents biographical sketches of archaeologists, and offers an overview of the prehistoric cultures which have occupied the Bay for about 5,000 years. Klein also includes several previously unpublished radiocarbon dates along with contemporary photographs of artifacts collected in 1883.

96 pages, 41 black/white photographs, over 50 maps and charts 5 1/2" x 8 1/2", index.
4,000 copies, ISBN 0-9651157-0-4
$12.95 plus postage

FOG ON THE MOUNTAIN
by Frederica de Laguna

Frederica de Laguna delights murder mystery fans with her cast of unusual characters in this well contrived tale set in Kachemak Bay. Originally published in 1938 for the Crime Club by Doubleday, Doran & Company, Fog on the Mountain is set in Southcentral Alaska where de Laguna conducted archaeological research in the early 1930's.

The 1995 soft cover reprint includes a new cover design and arrangement of the introductory pages including a Preface written by de Laguna in 1994.

288 pages, 4 maps, 5"x 7"
2,000 copies, Library of Congress #95-75337
$14.95 plus postage

ARCHAEOLOGY OF KACHEMAK BAY, ALASKA and *FOG ON THE MOUNTAIN*
are available from:
Kachemak Country Publications
P.O. Box 3406
Homer, AK 99603-3406
Telephone: (907) 235-8925